Chicago Public Library
Y0-AQX-174
Winter science projects.

Winter

Science Projects

JOHN WILLIAMS

Julian Messner
Parsippany, New Jersey

First published in 1995 by Evans Brothers Limited
2A Portman Mansions
Chiltern Street
London W1M 1LE
England

First published in 1997 in the United States by Julian Messner

A Division of Simon & Schuster
299 Jefferson Road
Parsippany, New Jersey 07054-0480

ISBN 0-382-39705-3 (LSB) 10 9 8 7 6 5 4 3 2 1
ISBN 0-382-39706-1 (PBK) 10 9 8 7 6 5 4 3 2 1

Acknowledgments
Planning and production by The Creative Publishing Company
Edited by Christine Lawrie
Designed by Ian Winton
Commissioned photography by Chris Fairclough
Illustrations by Annabel Spenceley
The publishers would like to thank the staff and pupils of East Oxford First School for their help in the preparation of this book.

For permission to reproduce copyright material, the author and publishers gratefully acknowledge the following: Action Plus Photographic: 20 (top); Bryan and Cherry Alexander: 22 (both); Bruce Coleman: (Francisco J Erize) 10 (left), (Norman Owen Tomalin) 19 (right), (Johnny Johnson) 23 (bottom), (Keith Gunnar) 25, (Bob Glover) 26 (bottom left), (Stephen J Kraseman) 26 (top), (Udo Hirsch) 27 (left); Econ Engineering Ltd: 12; Chris Fairlclough Colour Library: 6, 14, 15; Oxford Scientific Films (Richard Packwood) 16, (M Andera) 26 (bottom right).

Library of Congress Cataloging-in-Publication Data

Williams, John, 1939 Jan. 11--
Winter science projects/by John Williams
p. cm. (Seasonal science projects)
Originally published: London, England: Evans Brothers Ltd., 1996.
Includes index.
Summary: Presents a variety of projects and experiments appropriate to winter, including making a thermometer, melting ice, testing insulation, and making snowshoes. Includes notes for parents and teachers.
1. Science projects-- Juvenile literature. 2. Winter--Juvenile literature. [1. Science projects. 2. Science--Experiments. 3. Experiments. 4. Winter.] I. Title. II. Series: Williams, John, 1939 Jan. 11- Seasonal science projects.
Q182.3.W56 1996 96-17638
507'.8--dc20 CIP
AC

Contents

* Words in **bold** in the text are explained in the glossary.

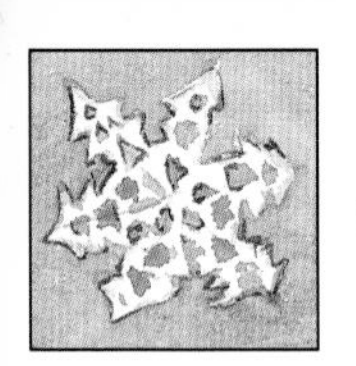

What is winter?

There are four seasons—spring, summer, autumn, and winter. Winter is the coldest season of the year. At the North and South Poles, the winter is long, dark, and very cold. It is not like this everywhere. In countries near the middle part of the world, winter may not be very different from summer.

Here are two signs that winter has come.

- The leaves have fallen from most of the trees.
- It's often cold or frosty in the morning.

Can you think of any more?

The trees in this picture have shed their leaves for the winter. What are trees that do this called?

PROJECT: Make a seasons zigzag book

The northern and southern parts of the world are called **hemispheres.** They have their winter at opposite times of the year. This zigzag book shows the seasons in each part.

You will need

- stiff paper
- Felt-tip pens

What to do

1. Fold the paper into four equal parts, like a zigzag.

2. Think of some simple pictures, or symbols, for each season. Draw a symbol on each part of the book, in the right order, starting with winter.

3. Turn the paper so the drawings are upside down. Then turn it over, keeping them upside down.

4. Draw the symbols again, but this time start with summer.

5. Write the months for the seasons of the Northern Hemisphere on one side of the book. Write the months for the seasons of the Southern Hemisphere on the other side.

Hint: When it's winter in the Northern Hemisphere, it's summer in the Southern Hemisphere.

Even though the seasons are different, the months will be on the same page on each side of your book.

Temperature

It's easy to tell if things are warm or cold by feeling them. Sometimes it's important to know exactly how warm or cold something is. For this we use an instrument called a thermometer. This measures temperature. When it is hot, the temperature is high. When it is cold, the temperature is low.

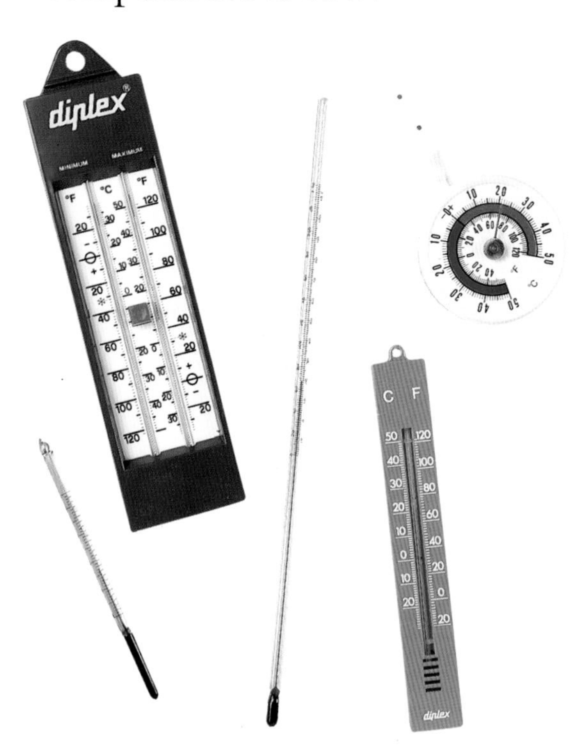

This picture shows some of the different kinds of thermometers that there are.

PROJECT: Make a thermometer

Temperature is measured in degrees. Not everybody uses the same measure. Some people use the **Fahrenheit scale**; some use the **Celsius scale**. You can make a thermometer showing both scales.

You will need

- Three pieces of stiff paper, two 8 in x 12 in (21 cm x 30 cm), one 6 in x 12 in (15 cm x 30 cm)
- Glue or stapler
- Felt-tip pens
- A real thermometer
- Scissors

What to do

1. Cut a long, thin slit in one large piece of stiff paper.

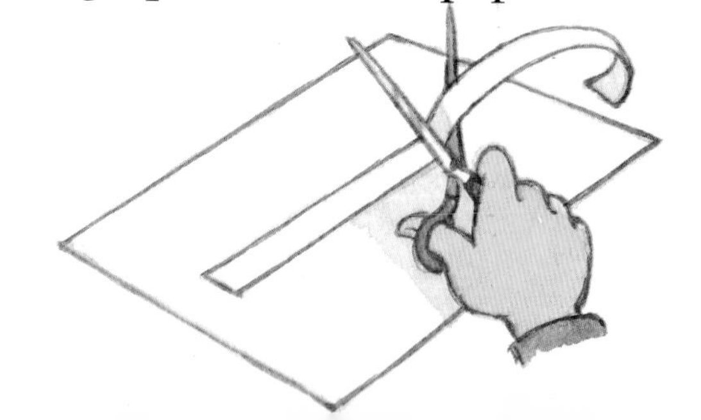

2. Glue or staple this piece of paper to the other large piece of paper along both long sides and one short side.

3. Color the middle of the narrow piece of paper red, and slide it between the other two pieces. The red color will show through the slit.

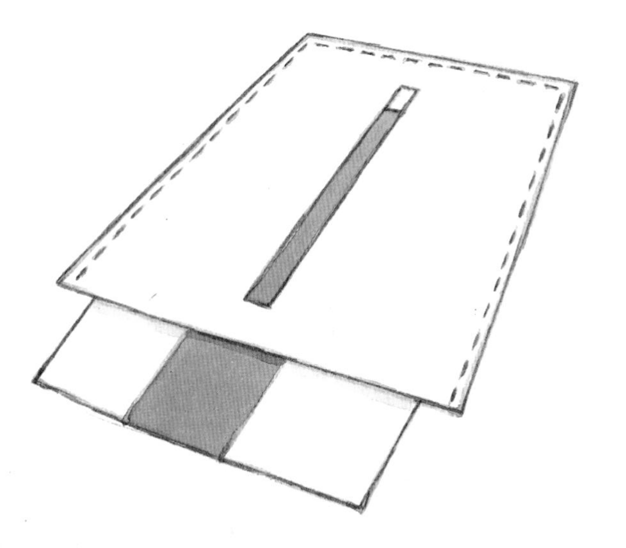

4. Copy the temperatures on this drawing onto your thermometer.

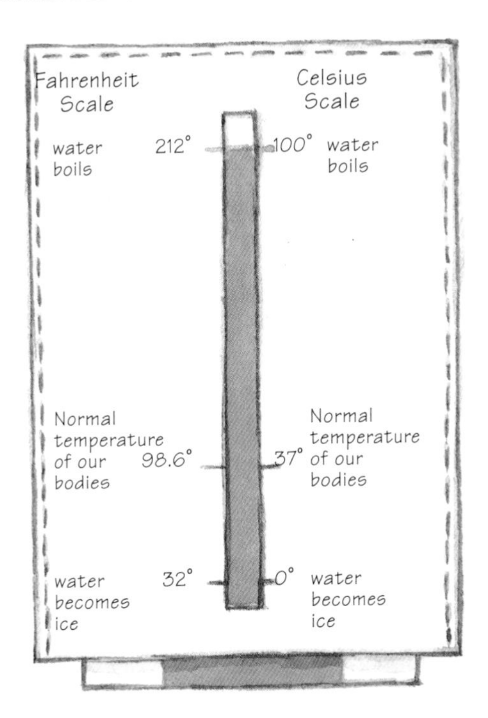

Take the temperature outside every day for a week with a real thermometer. Record it on your thermometer. Compare it to the temperature of your classroom or bedroom.

NEVER TOUCH ANYTHING YOU THINK MAY BE VERY HOT OR VERY COLD

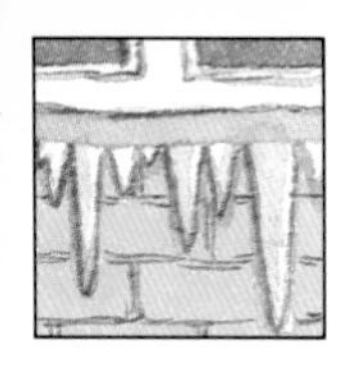

Looking at ice

When it is cold outside, water freezes on ponds and puddles. These **experiments** show you what happens to water when it freezes.

PROJECT: A freezing experiment

You will need

- A clear plastic cup, half filled with water
- A marker pen

What to do

1. Put a mark on the cup at the level of the water.

2. Place the cup in the freezer, until the water is frozen.

3. Look at the level of the ice in the cup. Has the level risen or fallen? Do you think there is any more or less water in the cup now that it is frozen?

4. Let the water melt again. Now where is the water level?

PROJECT: An iceberg experiment

Icebergs are huge masses of ice that float in the cold seas of the far North and far South. They can be a danger to ships. This project shows you why.

This iceberg is in the sea off the coast of Antarctica.

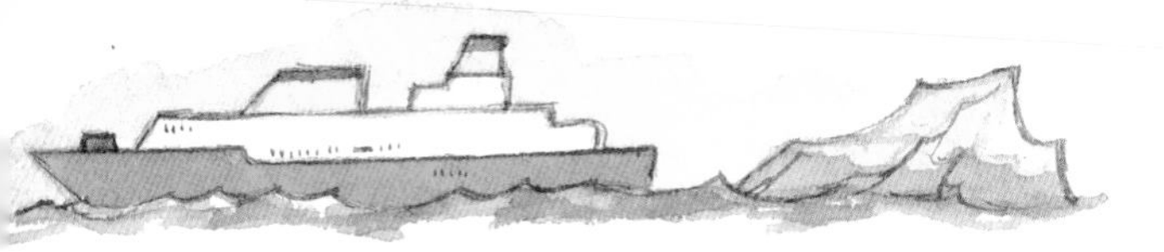

You will need
- Some ice cubes
- A deep tray of water
- A thermometer

What to do

1. Test the temperature of the water with a thermometer. Record it on a graph like the one on this page.

2. Float the ice cubes in the water. How much of the cube shows above the water? How much is under the water?

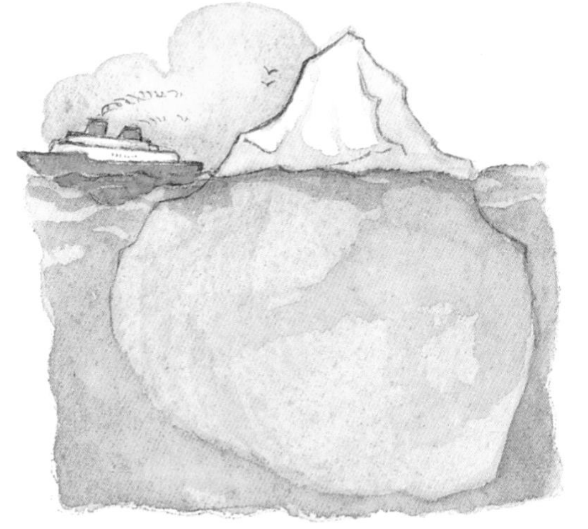

Much more of an iceberg is below the water than above it. This makes it a danger to ships.

3. Leave the ice cubes in the water until they disappear. Does the water feel colder or hotter as the ice cubes get smaller? Take the temperature of the water every 15 minutes as the ice melts.

4. Plot the temperature readings on your graph.

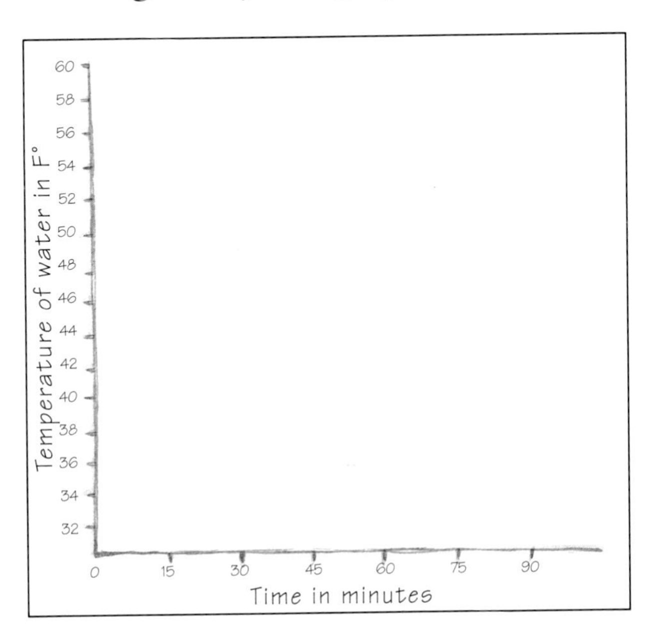

Melting ice

Local and state governments spread salt on roads in winter. This project will explain why they do this.

This truck is spreading salt on icy roads. It has big wheels to help keep it from skidding on the slippery ice and snow.

PROJECT: Ice and salt

You will need
- An ice cube
- Some salt

What to do

1. Sprinkle some salt onto a freshly frozen ice cube.

2. Describe what happens to the ice. Can you hear anything when the salt mixes with the ice?

It is more difficult for salty water to freeze than for fresh water. Salty water needs a lower temperature to turn into ice. This means it takes longer to freeze.

PROJECT: Another melting experiment

You will need

- A large block of ice (you can make one by freezing water in a plastic freezer carton
- A piece of string or wire
- Two 1 lb (0.5 kg) weights
- A shallow tray

What to do

1. Put your ice block on the tray. Wear gloves if you need to touch the ice.

2. Tie a weight to each end of the string or wire.

3. Put the piece of string or wire over the ice block so that it stretches tightly over the middle of the block. If your string or wire is very long, you can stand the tray on a box so the weights can hang freely.

4. What is happening to the wire or string? Look carefully at the ice. What is happening to the top of the block?

Try putting your ice block in the freezer. Does the same thing happen?

Weathering

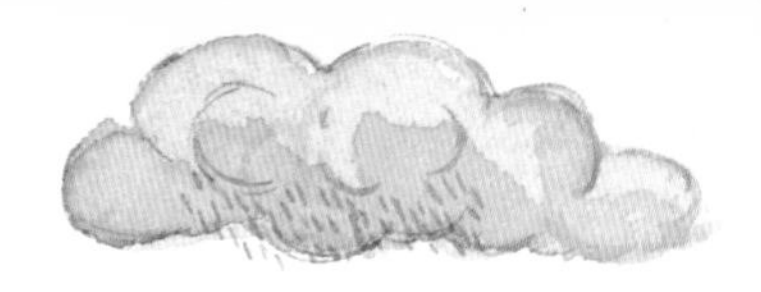

You have already seen that water takes up more space when it freezes. Water collects in the cracks of rocks. When it freezes, it breaks the rock open. Wind and rain wear the rocks away, too, but freezing will break them up first. This process is called weathering.

The small rocks have been broken off the mountain by ice.

Here are some experiments to show how ice breaks up rocks and soil.

PROJECT: Freezing and weathering

You will need

- Some soft rocks. Chalk is the best rock for this test, but if you can't get it, bits of broken house brick can be used instead.
- A shallow tray

What to do

1. Soak your rocks in water for two days.

2. Put them in the freezer, on the tray.

What happens to the rock when the water freezes?

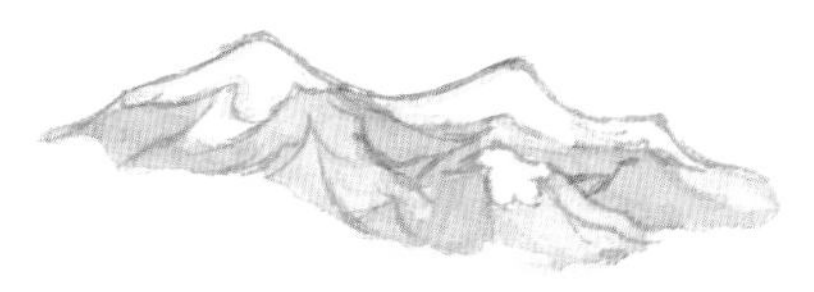
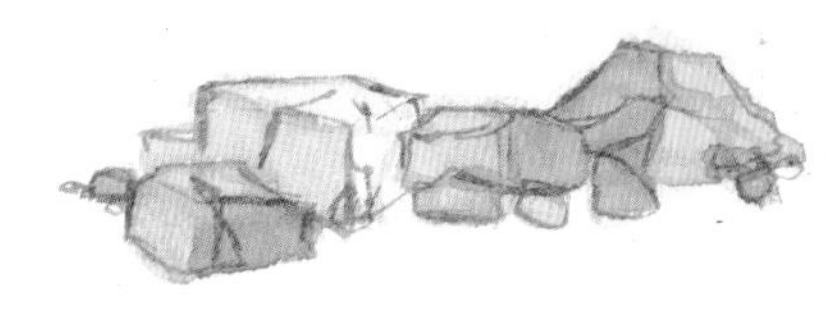

PROJECT: Frozen soil

Gardeners and farmers like the soil to be broken into small pieces. They dig or plow the soil in autumn, so that winter frosts can break up the lumps.

You will need

- Some clay from the garden, or try modeling clay
- A deep plastic tray

What to do

1. Put some wet clay in the tray.

2. Smooth it down carefully.

3. Put the tray in the freezer. Leave it at least a day.

How hard is the frozen clay?

4. Let the ice melt and see what difference it makes to the clay.

Keeping warm

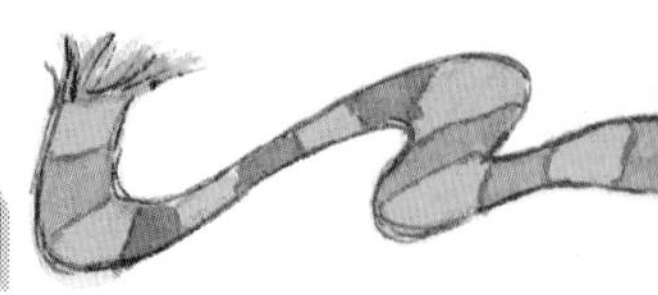

We all need to keep warm in cold weather. We wear warm clothes. Birds fluff up their feathers and trap warm air under them. Some animals have extra layers of fat or grow thicker fur.

This blue tit looks fat because it has fluffed out its feathers to help it keep warm.

We need to keep our houses warm. We light the fire or turn on the central heating. We put materials in the attic to keep the warm air inside. These act as **insulators**.

PROJECT: An insulation experiment

Some materials are better at insulating than others. The next experiment shows us how to find out which materials are the best.

ASK AN ADULT TO HELP WITH THIS EXPERIMENT

You will need

- Several cans, all the same size
- A box for each can, all the same size
- A thermometer for each can
- Aluminum foil
- Hot water
- Different materials. Try a woolen jumper or scarf, newspaper, sawdust, straw, or polystyrene chips.

What other things can you think of?

What to do

1. Put a can into each box.

2. Pack a different material round each can, leaving one with nothing around it.

3. Fill all the cans with hot water.

4. Make a lid for every can out of aluminum foil.

5. Measure the temperature of each can every five minutes for half an hour with your thermometers.

Write down your results on a chart like this one.

Time in minutes	Temperature in F°				
	Can 1	Can 2	Can 3	Can 4	Can 5
start 0					
5					
10					
15					
20					
25					
30					

Which material is the best insulator?
Why did we keep one can with nothing round it?
Insulating materials have many uses. Look around your home or school and see how many you can find.

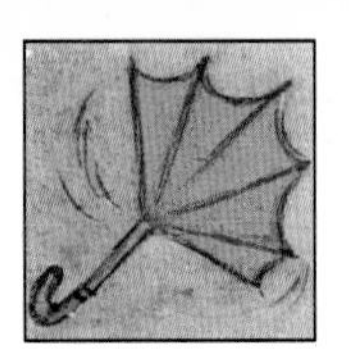

Windchill

We can keep warm by wearing extra clothes. When there is a cold wind, these may not be enough to protect you from **windchill**. Here is an experiment to find out what happens to the temperature if the wind is blowing.

PROJECT: Measuring windchill

You will need

- A large piece of paper
- A pencil
- A thermometer

What to do

1. Draw a plan of your school grounds, garden, or local park like the one at the bottom of this page.

2. Mark on it about ten places that are sheltered, but not under cover, for example, behind a bush or a tree.

3. Go outside when the wind is blowing. Measure the temperature in these sheltered places.

4. Take the temperatures again, standing as near to the sheltered places as you can, but this time in the wind. Are there any differences?

5. Write down your results like this.

place	Temperature in F°	
	out of wind	in wind
1. Behind tree		
2. Near bush		
3.		

In an experiment like this, the number of measurements is important.

You could measure just twice—once in the middle of the playground, garden, or park, and once against the side of a building. How might having only two sets of information, instead of ten, affect your results? Apart from being sheltered, can you think of other reasons why it might be warmer against the side of a building?

Clothes that protect you from windchill need not be thick, as long as the wind cannot blow through them. Skiing clothes are a good example of this.

Slipping and sliding

Above: This boy's ice skates are designed to slide easily on the ice. How are they different from ordinary shoes?

It is difficult to walk on ice as it is very smooth and you may slip, although it can be fun to slide or skate on it.

PROJECT: Testing shoes for grip

When we are just walking about, we don't want to slip and slide. Some shoes have ridges or patterns on the soles to keep us from sliding. Others don't. Look at your own and your friends' or family's shoes. Find out who is most likely to slip and fall!

You will need

- Several shoes of different types
- A 2 lb (1 kg) weight
- A spring balance

What to do

1. Put the weight inside a shoe.

2. Place the shoe on a smooth floor.

3. Hook the spring balance onto the top of the shoe.

4. Hold the spring balance and use it to pull the shoe along the floor.

How hard did you have to pull? The number shown on the scale of the spring balance is your measurement of pull. The higher this number is, the less likely you are to slip over in those shoes. Do this for all the shoes and write down your results.

type of shoe	measurement of pull		
	with shoe on the floor	with shoe on the carpet	with shoe on a sanded board
sneaker			
rain boot			

5. Now try doing the same experiment on different surfaces. Try carpet or a sanded board. Compare your results with your first set.

When there is ice on roads and pavements, local and state governments often put down sand or grit. Why do you think this is? What could they mix with it to make it work even better?

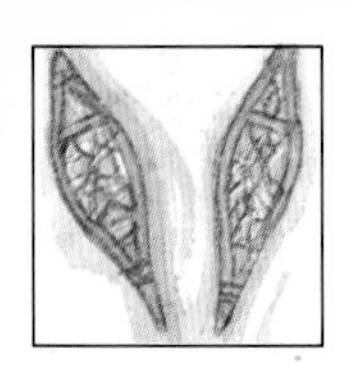

Snowshoes

Have you ever tried walking in deep snow? Imagine what it would be like walking through a snowdrift that reached up to your waist!

The Cree live in a part of North America where the land is often covered in deep snow. They wear special shoes, made of wood and strips of animal skin. This experiment will show you how these snowshoes work.

PROJECT: Make some snowshoes

You will need

- Toothpicks
- Modeling clay
- Two pieces of thick card, each about 1 in x 1 in (2 cm x 2 cm)
- Glue
- Some soft snow or very wet sand in a tray
- Scales and weights

These pictures show you what the snowshoes worn by the Cree look like.

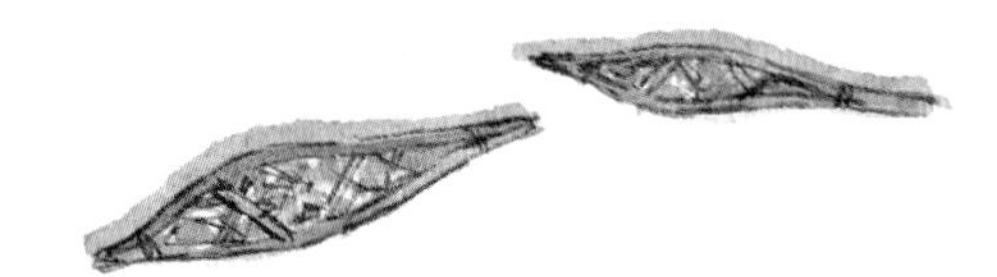

What to do

If you are using snow, do this experiment outside, or only bring it in when you are ready to use it.

1. Divide the modeling clay into two large balls weighing about 2 oz (60 g) and two smaller balls weighing about 0.5 oz (15 g).

2. Make two model people, using matchsticks for legs and arms.

3. Glue the two small squares of card to the feet of one of the models. You may need to use a little more clay to help the card stick.

4. Weigh the models again to check that they weigh the same. Add clay to one if it is lighter.

5. Lower them both gently onto the snow or wet sand. What happens to the model without the card shoes?

This polar bear has very wide feet, which help stop it from sinking into the snow.

Plants

Most plants stop growing in the winter. Soft leaves and stems that grew in summer die and fall off. Water, which plants need to grow, is often frozen solid. When it freezes, the water inside plants turns to ice, too. What would happen to a plant with a lot of water in its leaves and stems? Try this experiment, using your freezer.

PROJECT: Freezing plants

You will need

- Two raw potatoes
- A bowl
- A knife

What to do

1. Put one potato into the freezer for at least a day.

2. Take it out of the freezer and put it into the bowl. Let it **thaw** out.

3. Compare your thawed potato with the one that has not been frozen.

Try cutting them both in half.

ASK AN ADULT TO HELP WITH THIS EXPERIMENT

Water in plants is held in small units called **cells**. When the water freezes, it gets bigger and breaks the soft walls of these cells.

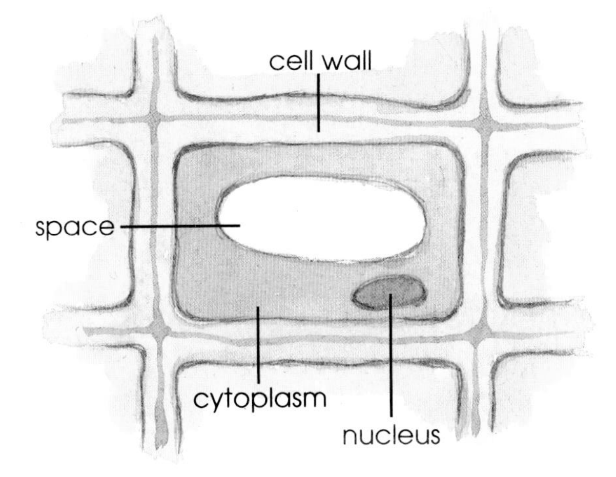

This is a simplified drawing of a plant cell. Cytoplasm is mostly made of water, which keeps the plant alive.

Some trees and plants are protected from cold winters. Look at the leaves of a pine tree or a holly tree, which stay green all winter. Is it easy or difficult to break or tear them? Why do you think they are shiny? Do the last experiment again with a handful of pine needles or holly leaves and some soft leaves. Get permission to use one or two soft leaves from an indoor plant if there are none outside. This time put them into plastic bags before putting them into the freezer. Compare your bags of leaves after they have thawed.

Pine trees can be covered in snow for many weeks in winter.

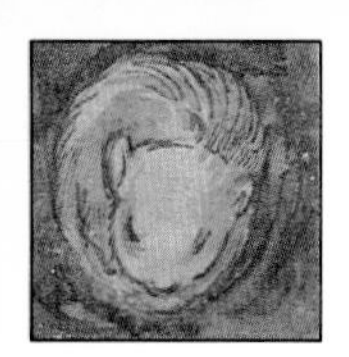

Animals

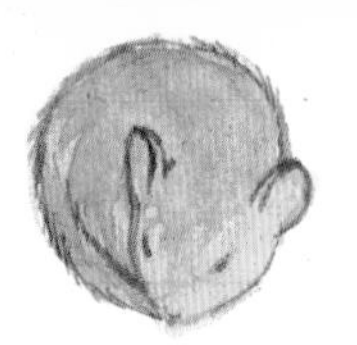

Wild animals find it difficult to get enough food in winter. The days are short, plants do not grow, and the ground may be covered in snow. The way that their bodies use up, or digest, food makes them warm. Thick fur or hair keeps the warmth in.

Animals survive in different ways. Some, like this squirrel, make stores of food, which they dig up when they are hungry.

Some birds and animals, like these caribou, move to warmer places for the winter. This is called migration.

Some animals, like this dormouse, sleep during winter. This is called hibernation.

Small animals have problems, because they get cold more quickly than larger animals.

This shrew has to keep eating for most of the day to keep warm. A day's food can weigh four times its own weight. It is so small that it will get cold very quickly if it does not eat all this food, and could die.

PROJECT: Food and warmth

You will need

- Two cans, one large and one very small
- Hot water
- Aluminum foil

In this experiment you have to imagine that the hot water is the can's food!

What to do

1. Fill the cans with hot water.

2. Make a lid for each can from aluminum foil.

3. Feel the cans with your hands every five minutes.

Which can gets cold first? Imagine the coldness means all the can's food is used up. You would have to refill it with hot water to keep it warm. This what the shrew has to do.

ASK AN ADULT TO HELP WITH THIS EXPERIMENT

Notes for teachers and parents

Day length

This topic is not included in the main text because of the difficulties for children in obtaining precise measurements, although they will know by direct observation that day length is shorter in winter. However, control technology, using a light-sensitive cell in conjunction with a suitable computer program, can be utilized to provide these measurements. The problem of continuous computer use for schools can be overcome by running the program on weekends, or even during the winter holiday. Children can also be made aware of the different day lengths, and the winter and summer months of both hemispheres, by the use of a globe. If a flashlight, representing the sun, is shone onto the globe, the resulting shadows will help show children the differences between the seasons, and, if the globe is spun, day and night as well.

Temperature

Perhaps we notice temperature more often in winter than any other time of the year. The extremes of coldness and warmth seem more obvious to us, whether it is through the necessity of wearing extra clothes or checking the level of the central heating. Making a thermometer will help children understand specific temperatures in the natural world. While they will already know the difference between hot and cold, they will not have experience of a range of temperature, their own body heat compared to the temperature of a warm bath, or just how cold it has to be for water to freeze. It is important that they should understand that a thermometer is just another measuring instrument. There is nothing magical about it. A thermometer measures degrees of temperature, just as a ruler measures centimeters or inches, or a clock measures time.

Water and ice

Most substances can exist as liquids, solids, or gases, or even as materials somewhere between these three states. However, water is the only one that children can easily experience in all three states. Water is at its most dense at 39°F (4°C), but freezes solid at 32°F (0°C). Most other substances are at their greatest density when solid. Water is different in that, when it freezes, its molecules reform into a lightweight lattice, and hence it will float. Children should not be expected to understand this in detail, but they might ask questions that will need some scientific answer.

The various experiments with ice in this section are designed to give children experience of experimental skills, such as observation and hypothesizing. History records that Michael Faraday carried out a similar series of experiments, although he would have understood less than we do about the nature and behavior of ice. When we float ice in water, the ice takes heat from the water. Eventually, equilibrium is reached when the ice has melted, and we have a dish of cold water. When we add salt to snow or ice, the salt dissolves in the surface layer of moisture, releasing heat, which melts the rest of the ice. The experiment with the large block of ice shows how the friction of the wire will generate heat, and so, temporarily at least, melt the ice. A similar process goes on between the blades of iceskates and the surface of the ice.

Weathering

Ice can affect the shape of the land in two ways. As great sheets of ice, or as large glaciers, it can carve out valleys and mountains. Even the hardest rocks can be ground down and reshaped by the forces exerted by moving ice. Although less dramatic, but in the long run just as effective and more widespread, ice will shape the landscape in the form of frost. The experiments in this section will help children understand the force exerted by the ice as it expands, breaking up rocks and soil, as well as affecting roads and buildings in a similar manner.

This is a good opportunity for children to understand the problems of frozen water pipes, and also car radiators. They can learn that water pipes need to be covered with a suitable material, to stop the water from freezing. They will understand that this cannot be done to a car radiator, but may know that antifreeze is mixed with the water, which affects the temperature at which it freezes.

Keeping warm

Various materials make good insulators, in that they do not conduct the heat away. Children should be encouraged to experiment with a range of materials, even if some of them are not apparently suitable. They could try water itself as an insulator. This set of experiments is valuable for showing what a "fair test" means. Variables can easily be excluded, and one can left with no insulation as a control, so that it can be seen that the can itself does not provide the protection.

Windchill

This section allows the children to see the effects of a cold wind on the temperature. Research has shown us that we are much more likely to suffer the effects of exposure if we are wet and left unprotected from cold wind. You can extend the experiment and simulate the effect of wet clothes by placing a small piece of wet tissue or cloth over the bulb of the thermometer. When we are wet, we lose heat quickly for two reasons. Water is a better conductor of heat than air, so if we have wet clothes and skin, heat is conducted away from our bodies more quickly than if we were dry. Secondly, the very process by which our wet clothes and skin dry out, evaporation, itself takes heat away from the body.

It is important to take as many different readings as possible. This will not only allow for the possibility of experimental error, but will also show if the difference in the strength of the wind has any effect on the temperature. It will also help to give a range of more accurate readings in the sheltered areas. We do not want the readings to be altered by the warmth from the building. The structure of the building itself can be a reservoir of warm air, and even warm air from an open door can affect the temperature. If we can avoid these variables, we will have a truer picture of what happens in nature. Animals trap a layer of air under their fur or feathers, which is warmed by their own body heat. Children need to be made aware that modern cold weather clothing is designed to do the same thing.

Slipping and sliding

Friction is a force. Forces are pushes and pulls, and friction tends to pull objects away from the normal direction that they wish to follow. We often want to reduce friction; for instance, we put oil into a car engine to reduce the wear and tear of constant friction. Walking on ice is an example of where increased friction is an advantage. With normal walking we rely on friction to allow us to push ourselves forward. Children can carry out experiments with different footwear on various surfaces from shiny to very rough to see which combinations give the best results. They might like to consider the implications of this in relation to tap dancing and running shoes.

Snowshoes

This experiment shows the relationship between weight, surface area, and pressure. It is very important that both the models are the same weight so that the effects of the card can be clearly seen. Where there are permanent environmental extremes, such as desert, extremes of temperature, or continuous snow and ice, people have to adapt their lives to these conditions. Wearing snowshoes is one such adaptation. Wearing special clothes or eating a particular diet are other ways by which people live in such conditions. We have seen in the sections on temperature and on windchill how humans can deal with these extremes. Animals and plants have had to evolve, over many years, mechanisms to allow them to survive. These themes are expanded upon in the next two sections.

Plants

Plants protect themselves in winter in various ways. Some pass the winter as seeds, while other, larger plants shed their leaves and remain dormant. Some have evolved as evergreens and have leaves with a protective coating of wax, although even evergreens will remain dormant during a cold winter, especially if the soil water remains frozen. Plants that have not evolved these mechanisms would not survive a cold winter. When a green plant is affected by frost, the water in the plant freezes, and by expanding breaks the cell wall. It is difficult to explain this to young children. They will see the effect of the frost—a brown, wilted plant. They cannot see the damage inside the plant. One possible way to explain this process is to fill a small plastic bag with water, tie it tightly, making sure there is no free air in the bag, and let the children feel it. It will be firm but still pliable, and can represent some of the softer kinds of plant cell (most plant cells have rigid walls). Put the bag into the freezer, freeze, and then allow it to thaw. Hopefully the bag will have burst under the pressure of the ice, and the children can more easily imagine the effect this would have on a plant.

Animals

Animals have even more ways of surviving winter than plants. We have already seen how they can protect themselves from the cold, although it is often the shortage of food and water that causes problems. The experiment in this section shows how animals that have a large surface area in comparison to their volume will quickly lose heat energy. Some animals have evolved metabolic systems that allow them to go without food for several weeks, or to store food and water in their bodies in the form of fat. Some migrate, some hibernate. Children need to understand that hibernation is not just "going to sleep for the winter." Few warm-blooded animals truly hibernate. The process is even now not fully understood, although it involves the slowing down of all body processes, so that very little energy is required to stay alive. Hibernation is thought to be brought about by day length as much as temperature, although in cold-blooded animals temperature does have a more direct effect.

Glossary

Cell A tiny part of a plant or animal. Even a very small plant or animal can be made up of millions of cells.

Celsius scale A way of measuring temperature, named after Anders Celsius, which has 0° as the freezing point of water, and 100° as its boiling point.

Experiment A test made to discover something that is not known.

Fahrenheit scale A way of measuring temperature, which has 32° as the freezing point of water, and 212° as its boiling point.

Hemisphere A half of a ball (or sphere). The world is a sphere, and is divided into northern and southern halves, or eastern and western halves. Each half is called a hemisphere.

Insulator A material that can stop heat or electricity from getting through. Most things that are not metal are good insulators.

Thaw To raise the temperature of something, usually ice, so that it melts.

Windchill The extra coldness caused by the wind, often when blowing on wet clothes or skin.

Index